This Book Belongs To

Name :

Phone :

Address :

Project Name :

Foreman :

Project No :
Date :
Day :

Visitors

Schedule

Problems

Safety Issues

Summary Of Work

Signature :

Employee	Trade	Hours	Overtime

Equipment On Site	No. Units

Materials Delivered	No. Units	Equipment Rented	Rate

Others

Notes :

Project Name :

Foreman :

Project No :
Date :
Day :

Visitors

Schedule

Problems

Safety Issues

Summary Of Work

Signature :

Employee	Trade	Hours	Overtime

Equipment On Site	No. Units

Materials Delivered	No. Units	Equipment Rented	Rate

Others

Notes :

Project Name :

Foreman :

Project No :
Date :
Day :

Visitors

Schedule

Problems

Safety Issues

Summary Of Work

Signature :

Employee	Trade	Hours	Overtime

Equipment On Site	No. Units

Materials Delivered	No. Units	Equipment Rented	Rate

Others

Notes :

Project Name :

Foreman :

Project No :

Date :

Day :

Visitors

Schedule

Problems

Safety Issues

Summary Of Work

Signature :

Employee	Trade	Hours	Overtime

Equipment On Site	No. Units

Materials Delivered	No. Units	Equipment Rented	Rate

Others

Notes :

Project Name :

Foreman :

Project No :
Date :
Day :

Visitors

Schedule

Problems

Safety Issues

Summary Of Work

Signature :

Employee	Trade	Hours	Overtime

Equipment On Site	No. Units

Materials Delivered	No. Units	Equipment Rented	Rate

Others

Notes :

Project Name :

Foreman :

Project No :
Date :
Day :

Visitors

Schedule

Problems

Safety Issues

Summary Of Work

Signature :

Employee	Trade	Hours	Overtime

Equipment On Site	No. Units

Materials Delivered	No. Units	Equipment Rented	Rate

Others

Notes :

Project Name :

Foreman :

Project No :
Date :
Day :

Visitors

Schedule

Problems

Safety Issues

Summary Of Work

Signature :

Employee	Trade	Hours	Overtime

Equipment On Site	No. Units

Materials Delivered	No. Units	Equipment Rented	Rate

Others

Notes :

Project Name :

Foreman :

Project No :
Date :
Day :

Visitors

Schedule

Problems

Safety Issues

Summary Of Work

Signature :

Employee	Trade	Hours	Overtime

Equipment On Site	No. Units

Materials Delivered	No. Units	Equipment Rented	Rate

Others

Notes :

Project Name :

Foreman :

Project No :
Date :
Day :

Visitors

Schedule

Problems

Safety Issues

Summary Of Work

Signature :

Employee	Trade	Hours	Overtime

Equipment On Site	No. Units

Materials Delivered	No. Units	Equipment Rented	Rate

Others

Notes :

Project Name : ....

....

Foreman : ....

Project No :
Date :
Day :

Visitors

Schedule

Problems

Safety Issues

Summary Of Work

Signature : ....

Employee	Trade	Hours	Overtime

Equipment On Site	No. Units

Materials Delivered	No. Units	Equipment Rented	Rate

Others

Notes :

Project Name :

Foreman :

Project No :
Date :
Day :

Visitors

Schedule

Problems

Safety Issues

Summary Of Work

Signature :

Employee	Trade	Hours	Overtime

Equipment On Site	No. Units

Materials Delivered	No. Units	Equipment Rented	Rate

Others

Notes :

Project Name :

Foreman :

Project No :
Date :
Day :

Visitors

Schedule

Problems

Safety Issues

Summary Of Work

Signature :

Employee	Trade	Hours	Overtime

Equipment On Site	No. Units

Materials Delivered	No. Units	Equipment Rented	Rate

Others

Notes :

Project Name :

Foreman :

Project No :
Date :
Day :

Visitors

Schedule

Problems

Safety Issues

Summary Of Work

Signature :

Employee	Trade	Hours	Overtime

Equipment On Site	No. Units

Materials Delivered	No. Units	Equipment Rented	Rate

Others

Notes :

Project Name :

Foreman :

Project No :
Date :
Day :

Visitors

Schedule

Problems

Safety Issues

Summary Of Work

Signature :

Employee	Trade	Hours	Overtime

Equipment On Site	No. Units

Materials Delivered	No. Units	Equipment Rented	Rate

Others

Notes :

Project Name :

Foreman :

Project No :
Date :
Day :

Visitors

Schedule

Problems

Safety Issues

Summary Of Work

Signature :

Employee	Trade	Hours	Overtime

Equipment On Site	No. Units

Materials Delivered	No. Units	Equipment Rented	Rate

Others

Notes :

Project Name :

Foreman :

Project No :
Date :
Day :

Visitors

Schedule

Problems

Safety Issues

Summary Of Work

Signature :

Employee	Trade	Hours	Overtime

Equipment On Site	No. Units

Materials Delivered	No. Units	Equipment Rented	Rate

Others

Notes :

Project Name :

Foreman :

Project No :
Date :
Day :

Visitors

Schedule

Problems

Safety Issues

Summary Of Work

Signature :

Employee	Trade	Hours	Overtime

Equipment On Site	No. Units

Materials Delivered	No. Units	Equipment Rented	Rate

Others

Notes :

Project Name :

Foreman :

Project No :
Date :
Day :

Visitors

Schedule

Problems

Safety Issues

Summary Of Work

Signature :

Employee	Trade	Hours	Overtime

Equipment On Site	No. Units

Materials Delivered	No. Units	Equipment Rented	Rate

Others

Notes :

Project Name :

Foreman :

Project No :
Date :
Day :

Visitors

Schedule

Problems

Safety Issues

Summary Of Work

Signature :

Employee	Trade	Hours	Overtime

Equipment On Site	No. Units

Materials Delivered	No. Units	Equipment Rented	Rate

Others

Notes :

Project Name :

Foreman :

Project No :
Date :
Day :

Visitors

Schedule

Problems

Safety Issues

Summary Of Work

Signature :

Employee	Trade	Hours	Overtime

Equipment On Site	No. Units

Materials Delivered	No. Units	Equipment Rented	Rate

Others

Notes :

Project Name :

Foreman :

Project No :
Date :
Day :

Visitors

Schedule

Problems

Safety Issues

Summary Of Work

Signature :

Employee	Trade	Hours	Overtime

Equipment On Site	No. Units

Materials Delivered	No. Units	Equipment Rented	Rate

Others

Notes :

Project Name :

Foreman :

Project No :
Date :
Day :

Visitors

Schedule

Problems

Safety Issues

Summary Of Work

Signature :

Employee	Trade	Hours	Overtime

Equipment On Site	No. Units

Materials Delivered	No. Units	Equipment Rented	Rate

Others

Notes :

Project Name :

Foreman :

Project No :
Date :
Day :

Visitors

Schedule

Problems

Safety Issues

Summary Of Work

Signature :

Employee	Trade	Hours	Overtime

Equipment On Site	No. Units

Materials Delivered	No. Units	Equipment Rented	Rate

Others

Notes :

Project Name :

Foreman :

Project No :
Date :
Day :

Visitors	Schedule

Problems	Safety Issues

Summary Of Work

Signature :

Employee	Trade	Hours	Overtime

Equipment On Site	No. Units

Materials Delivered	No. Units	Equipment Rented	Rate

Others

Notes :

Project Name :

Foreman :

Project No :
Date :
Day :

Visitors

Schedule

Problems

Safety Issues

Summary Of Work

Signature :

Employee	Trade	Hours	Overtime

Equipment On Site	No. Units

Materials Delivered	No. Units	Equipment Rented	Rate

Others

Notes :

Project Name :

Foreman :

Project No :
Date :
Day :

Visitors

Schedule

Problems

Safety Issues

Summary Of Work

Signature :

Employee	Trade	Hours	Overtime

Equipment On Site	No. Units

Materials Delivered	No. Units	Equipment Rented	Rate

Others

Notes :

Project Name :

Foreman :

Project No :
Date :
Day :

Visitors

Schedule

Problems

Safety Issues

Summary Of Work

Signature :

Employee	Trade	Hours	Overtime

Equipment On Site	No. Units

Materials Delivered	No. Units	Equipment Rented	Rate

Others

Notes :

Project Name :

Foreman :

Project No :
Date :
Day :

Visitors

Schedule

Problems

Safety Issues

Summary Of Work

Signature :

Employee	Trade	Hours	Overtime

Equipment On Site	No. Units

Materials Delivered	No. Units	Equipment Rented	Rate

Others

Notes :

Project Name :

Foreman :

Project No :
Date :
Day :

Visitors

Schedule

Problems

Safety Issues

Summary Of Work

Signature :

Employee	Trade	Hours	Overtime

Equipment On Site	No. Units

Materials Delivered	No. Units	Equipment Rented	Rate

Others

Notes :

Project Name :

Foreman :

Project No :
Date :
Day :

Visitors

Schedule

Problems

Safety Issues

Summary Of Work

Signature :

Employee	Trade	Hours	Overtime

Equipment On Site	No. Units

Materials Delivered	No. Units	Equipment Rented	Rate

CPSIA information can be obtained
at www.ICGtesting.com
Printed in the USA
LVHW051934130323
741532LV00014B/1645

9 783986 081089